SMALL ENGINES MAINTENANCE & REPAIR

Text by Calton E. Taylor
Introduction and Photography by Forest H. Belt

THEODORE AUDEL & CO.
a division of
HOWARD W. SAMS & CO., INC.
4300 West 62nd Street
Indianapolis, Indiana 46268

FIRST EDITION

FIRST PRINTING—1974

International Standard Book Number: 0-672-23800-4

PREFACE

If you own a lawnmower, weedcutter, snowblower, or any machine powered by a small gasoline engine, you probably know it needs a tune-up at least once every year. After you have owned it for awhile, you will discover it also needs repairs periodically.

More often than not, a good repairman is hard to find. There is no good reason for this, except that few mechanics seem to want to work on small gasoline engines—or the ones that do are very busy. Why most shy away from small engines is hard to say. Small engines are surprisingly simple. You can take one completely apart and put it back together in an hour or so—that is, if you know how. In the process, if you are aware of what to look for, you can spot and fix any defect that keeps the engine from starting or running properly.

As an example, suppose you can't get your small gasoline engine started. You have only two major areas of its operation to investigate: *carburetion* and *ignition.* Any trouble is likely to be in one or the other locations. If the carburetor sprays the right mixture of gasoline/air into the combustion chamber, and the spark plug introduces the right kind of spark at the right time, the engine has to start and run. Find out what fouls up either of these actions and you can fix the machine.

These two areas of troubleshooting occupy the major portion of this book. The *first section* deals with carburetion. You discover that the fastest way to find trouble in a carburetor is to take it apart and clean it. While it is torn down, you can inspect critical components. You need not be awed by the intricacy of a carburetor. Photos and lucid explanations take you step-by-step through disassembly and reassembly of the most popular types. You will find other carburetors on different models of small engines, but they fit the categories typified here.

A tune-up, in small-engine jargon, involves replacing the spark plug, breaker points, and condenser. The *second section* of this book leads you through complete procedures for tune-ups and ignition repairs on a number of engine types. You can see how to replace and adjust the ignition coil and to check the magnets that work with the coil to generate an ignition spark. As with carburetors, these ignition systems represent the most common varieties. You can apply the photos and instructions to other brands and models.

Finally, there is something you should realize about small engines. You will find a raft of brand name lawnmowers, composters, generators, saws, and so on. But they incorporate only a few kinds of engines. That makes it easy for you to learn to repair and maintain your own small gasoline engine. The examples throughout this book are typical of just about all kinds.

We hope you find our book valuable. We have taken great care with the text and photographs to be as accurate and revealing as possible. You should save a lot of money on repairs, and your small engines should last many years instead of just one or two seasons.

Forest H. Belt

CONTENTS

SECTION 1
Repairs In And Around Carburetors

SECTION 2
Tune-ups And Ignition Repairs

Repairs in and Around Carburetors

You find two basic types of carburetors on small gasoline engines. One is called *gravity-feed;* the other, *siphon-feed.*

In a gravity-feed carburetion system, the gas tank mounts separately from the carburetor and higher on the engine, as shown in Fig. 1. Gasoline transfers from the tank to the carburetor through a small rubber hose. The weight of the

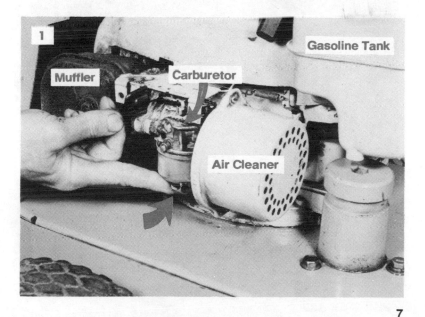

1

Gasoline Tank

Muffler   Carburetor

Air Cleaner

fuel exerts the only pressure that makes gasoline flow to the carburetor. A *float* and *needle valve* in the carburetor control the rate of flow, keeping the carburetor *bowl* full as the engine uses the fuel.

The gasoline tank in a siphon-feed system mounts directly beneath the carburetor, as shown in Fig. 2. It and the carburetor work together as one unit. A piece of plastic tubing from the carburetor almost touches the bottom of the fuel tank. Vacuum from the engine pulls the gasoline up into the carburetor. This type of carburetor has no float and needle valve to control fuel flow—they are not necessary. The engine vacuum lifts up only as much gasoline as the engine needs to keep running.

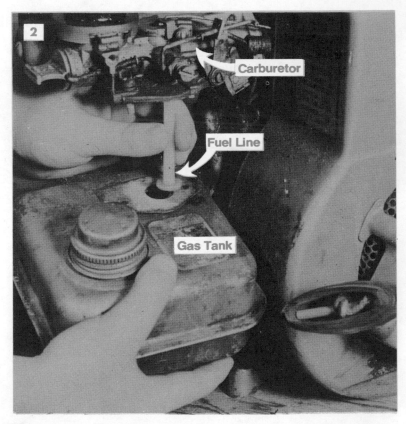

THE GRAVITY-FEED CARBURETOR

Give yourself easy access to the engine. Remove whatever hardware gets in the way. Almost all rider-type lawnmowers have a seat or part of its mounting bracket obstructing the engine. A couple of bolts usually frees the seat assembly.

Let's begin with checkout and minor repairs for a gravity-feed carburetor. You'll have to remove the air cleaner cover for access to the fuel line, where you'll do your first fuel-system checking. The air cleaner cover is held to the carburetor by a pair of screws. Most housings have twist-slots instead of holes, so you don't have to remove the screws completely. See Fig. 3.

Next, extract the air cleaner filter from the housing cup, as shown in Fig. 4. If it's very dirty, you will need to replace it. There is an easy method for checking this type of filter. Hold it up to the sun or to a bright light and try to look through it. If little or no light can be seen, the element is too dirty to clean the air properly.

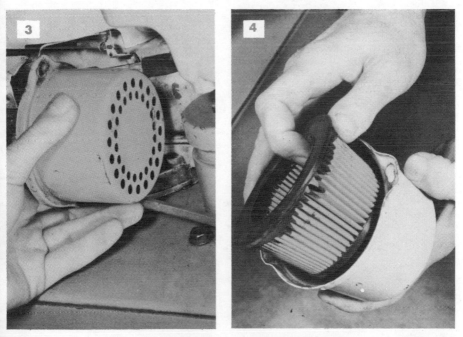

Now is the time to check the choke butterfly. It is a minor part, but vital to proper starting, especially on cold days. It mounts inside the air-intake throat of the carburetor, next to the end where the air cleaner attaches. See Fig. 5. When the choke knob, or lever, is in the open position, the butterfly should be turned so its edge cuts the incoming air. This allows air free passage into the carburetor. When the choke is fully closed, (Fig. 6) the butterfly should be turned exactly crosswise over the intake opening. This blocks off incoming air. The fuel-air mixture for the carburetor becomes much richer, making it easier to start the engine.

Make sure the choke lever and cable are positioned so the control marks coincide with actual movement of the butterfly. While you are looking at the choke butterfly, verify that the screws which hold the butterfly plate to its shaft have not vibrated loose.

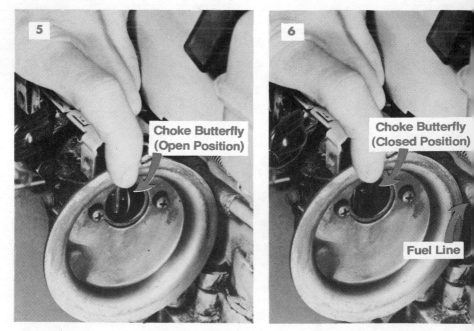

5 Choke Butterfly (Open Position)

6 Choke Butterfly (Closed Position)

Fuel Line

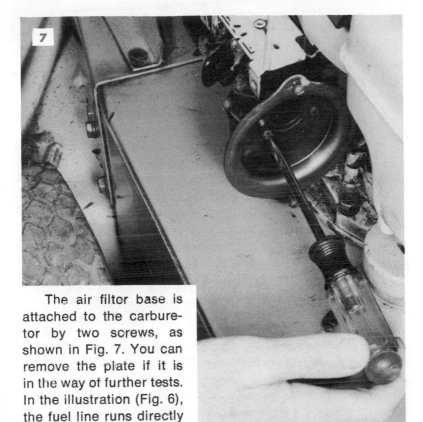

The air filter base is attached to the carburetor by two screws, as shown in Fig. 7. You can remove the plate if it is in the way of further tests. In the illustration (Fig. 6), the fuel line runs directly behind it.

Now you should check the fuel flow. Pull the fuel-line hose loose from the carburetor. The stream of gasoline from the tank should be as big in diameter as the inside of the fuel line. If it isn't, remove the gas tank and wash it out with clean gasoline. The object is to dislodge and flush out any sediment. Before you refill the tank, blow through the line itself (as illustrated in Fig. 8) to be sure it is not blocked.

Attach a 2-ft. length of clean hose to the carburetor fuel-inlet pipe. Blow through the hose into the carburetor. This should clear away any sediment that could stop up the spray jet that creates the air-fuel vapor, or that could stick in the seat of the needle valve (explained later). Use only your mouth to do this; compressed air is too powerful, the force of air might damage a jet or jet seat.

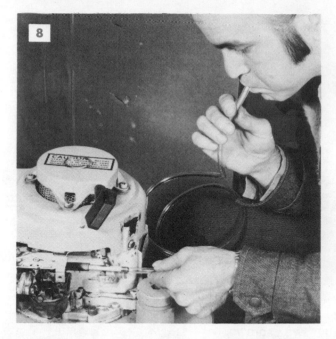

The next logical step, with this type of carburetor, is to check the float-bowl drain valve. The valve is on the bottom of the float bowl, usually toward the outer edge, as shown in Fig. 9. Whenever you push the knob in or sideways, gasoline should dribble out freely. This is done when the engine and tank are all assembled, to check gasoline flow from the gas tank to the carburetor bowl. The drain valve is spring loaded to hold it against its seat. About the only fault that can develop here is leakage past the seat. If this valve goes bad, the entire float bowl must be replaced.

The other knob (Fig. 9) on the bottom of the float bowl is the run-mixture screw. This screw controls the air-fuel mixture ratio at working speed. As with idle mixture, you never turn this screw unless the engine is running. *Do not* attempt to adjust any carburetor screw just because the engine will not start. This is almost never the answer.

A metal plate on top of the carburetor holds several carburetor controls. To gain access to the carburetor or to repair the controls, you may have to remove this plate. A pair of screws fastens it directly to the top of the carburetor.

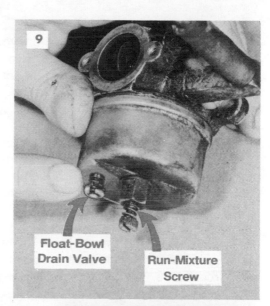

**Float-Bowl
Drain Valve**

**Run-Mixture
Screw**

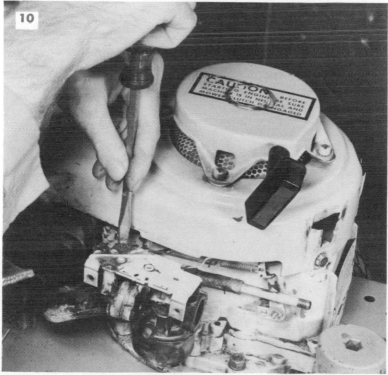

13

A wire runs from a tab on the cover plate to the engine's ignition magneto (Section 2). When you pull the throttle-control lever to the stop position, a movable tab touches the terminal attached to the magneto wire. The movable tab is riveted to the metal plate and therefore grounds the coil of the magneto, preventing any ignition sparking from reaching the spark plug. This kills the engine.

Clean the underside of this cover plate thoroughly with gasoline. Clean away all dirt and grease from the wire terminal and the throttle-control-cable tab to ensure good contact. A small rod from the carburetor is also connected to the throttle-control lever on the metal plate. This rod operates the throttle and is connected to the governor. The rod has an offset end that holds it in the control-lever hole. Make sure the rod is not bent and that it works the control lever freely. See Fig. 11 for the various parts mentioned.

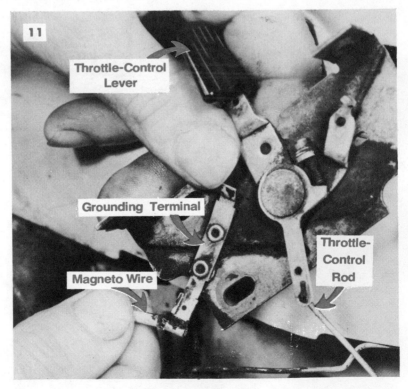

11

Throttle-Control Lever

Grounding Terminal

Magneto Wire

Throttle-Control Rod

Inspect the magneto kill wire thoroughly. See if the wire has frayed or come loose at the terminal. If the connection is bad, remove and replace the terminal. Check for cracks in the insulation over the whole length of the wire. Minor cracks can be covered by a piece of plastic tape, although it should be replaced with a new wire. Check the wire terminal at the grounding tab on the cover plate. The terminal should fit tightly on the tab, for a solid electrical connection. A tight connection assures that the kill system will work properly. After removing the wire from the tab and the throttle-control rod from the control lever, the cover plate is free to be removed from the engine.

After removing the carburetor plate, the carburetor can then be removed from the engine. Generally the bolts that hold the intake pipe to the engine come out first, as shown in Fig. 12. That's the only way you can get at the bolt on the back side. The carburetor is situated too close to the engine for a wrench to turn in the limited space.

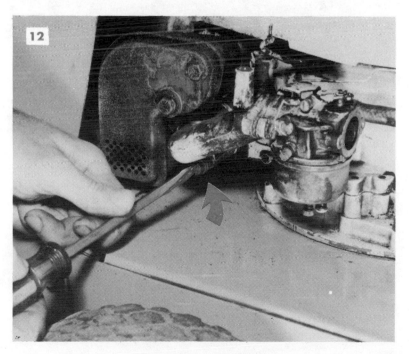

After you remove the bolts on the intake pipe flange, turn the carburetor and intake pipe around to allow access to the one bolt that can't be reached otherwise. Lay the carburetor on the mower housing while you are working on it. A cloth would be handy for this. Remove the two bolts holding carburetor to intake pipe, as shown in Fig. 13. Be careful that no dirt gets inside the carburetor. Any foreign material left inside can damage the carburetor. At best, it will keep the engine from running smoothly.

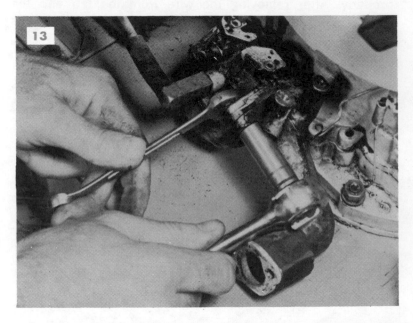

As part of a total carburetor inspection, remove the run-mixture needle valve and screw as shown in Fig. 14. *First,* however, (counting the number of turns) turn the screw clockwise until it just touches the seat. Later, when you replace the screw, you can set it at the same position as before. That way the engine will start easier and run while you readjust the mixture setting.

After you have removed the screw, see if a groove is worn in the tapered end where it contacts the seat. See Fig. 15. If you find much of a groove in the taper, or if the point of the needle is bent, replace the screw and seat valve assem-

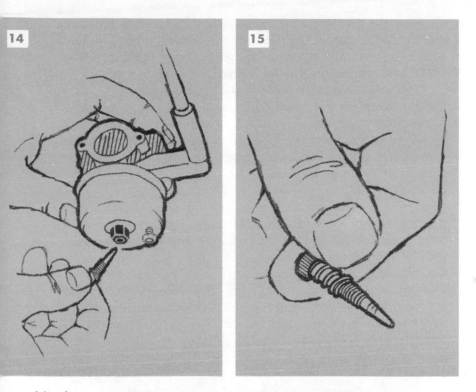

bly. A groove on the taper or a bent point signifies someone has forced the needle valve too tight against the seat. It could happen when you turn the screw inward to count the number of turns.

Now, here is a suggestion that seems contradictory to earlier warnings. If a new engine will not start when you first purchase it, and you believe someone has messed with the mixture screws, set the run-mixture by this method: Turn the needle-valve screw all the way in until it just touches the seat. Then back the screw out approximately one and one-half turns. This almost always puts it close enough to let you start the engine.

The body of the needle-valve seat for the run-mixture adjustment is made like a bolt, and it holds the float bowl in place on the carburetor. Wipe off the outside of the bowl, before you unscrew the seat from the carburetor. Any grease or dirt that gets inside the carburetor can ruin it. See Fig. 16.

After the carburetor is clean, you can remove the seat and screw as a unit. Do not turn the screw within the seat body, and you will not have to readjust the mixture screw drastically when you put the carburetor back together.

After you have the needle valve and seat out of the carburetor, examine the small holes in the outside of the seat. These holes must not be plugged by sediment or dirt, which would interrupt the flow of gasoline. See Fig. 17. Hold the seat up to the light and look through the holes. To be doubly sure, wash the seat in gasoline before you replace it in the carburetor.

If no light can be seen through the small holes, back the needle valve out of the seat body. Hold the end of a short plastic hose against the hexagon end of the seat and blow. It will not hurt the seat to blow it out with compressed air, but that much pressure usually is not necessary. If blowing with your mouth will not clear the holes, thread a length of fine wire through the holes to push out any debris. Then blow through the hose again. Remember to wash the seat in clean gasoline. See Fig. 18.

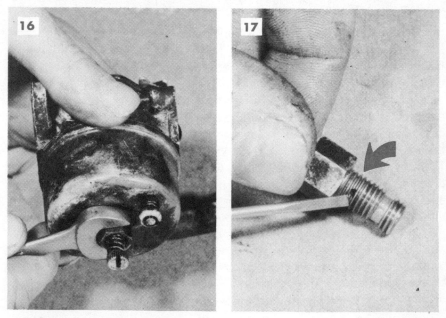

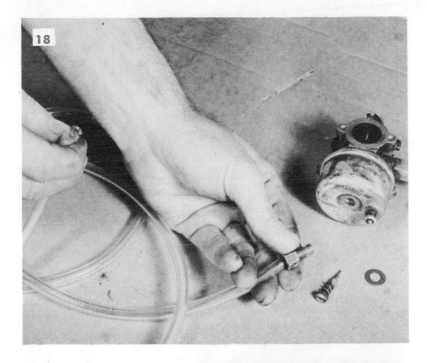

18

19

With the needle valve and seat removed, you can take the float bowl off, as shown in Fig. 19. Be extremely careful with the float assembly as it can be damaged easily. If the float arm is bent, the float cannot move freely and will fit loose on the pin. The engine will not start or will flood out and quit immediately.

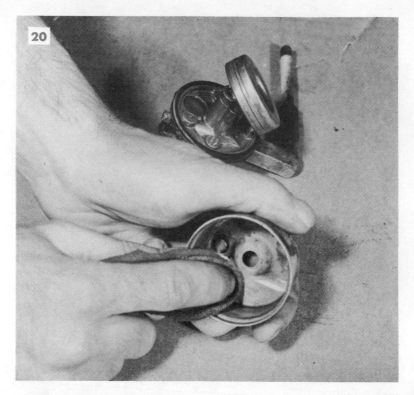

Wash out the inside of the float bowl thoroughly with gasoline, as shown in Fig. 20. This removes all the sediment that builds up in the bottom of the bowl. Sometimes the heat of the engine hardens sediment into a cake-like substance. You may have to scrape it free with a screwdriver or knife. Never reinstall the float bowl without first cleaning it.

An idle-mixture screw protrudes from the side of the carburetor just above the float bowl. When the carburetor is on the engine, you can reach the screw easily with a screwdriver to adjust it. You turn this screw *only* while the engine is running. This screw is shown in Fig. 21. A needle valve at the other end of this screw controls the air-fuel mixture ratio while the engine runs at idle speed. Never tamper with the mixture screws when the engine is not running. That is one of the most common mistakes an amateur makes when an engine will not start. Turning this adjustment screw will not help at all.

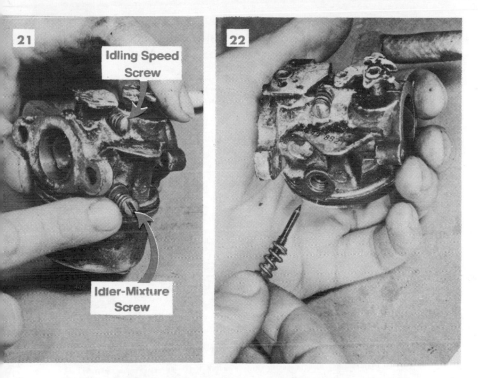

Another adjustment on the carburetor controls idling speed. See Fig. 21. This screw sits on top of the carburetor. It is an adjustable stop for the minimum setting of the throttle. You set it to make the engine run at as slow a speed as it can without skipping or trying to die out. You may jump back and forth between the idle speed and idle-mixture screws a couple of times to get smoothest idling, but only after you have the engine running.

While you have the carburetor dismantled, turn the idle-mixture needle-valve screw inward (clockwise) until it lightly touches the seat. Count the turns so you can return to the same setting.

Then back the needle valve out and inspect it for wear or damage, as shown in Fig. 22. The chief way it gets damaged is through being tightened too firmly against the seat. If you can see a groove on the tapered end, replace the needle valve, because the groove interferes with the flow of gas past the seat which renders an impossible setting for a proper mixture. Only a dark ring should be visible, where the taper fits next to the seat. This ring is only a discoloration of the metal caused by chemicals in the gasoline. See Fig. 23.

The carburetor float consists of a hollow, airtight metal donut pivoted so it works a needle valve to control the flow of gasoline from the tank into the carburetor bowl. The float assembly is shown in Fig. 24. A pivot pin (Fig. 25) mounts the float to the carburetor by two posts. The float must pivot freely on the pin.

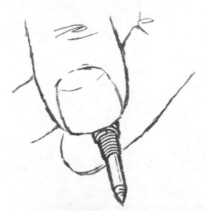

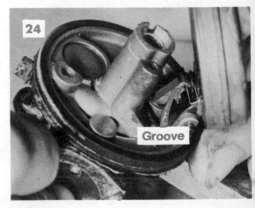

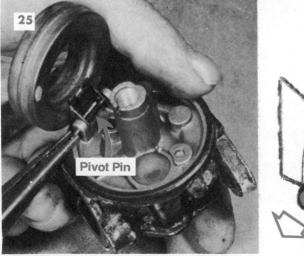

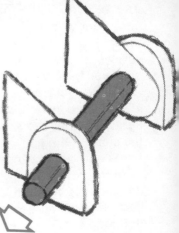

The groove in the float needle valve, (Fig. 26) fits into a small slot in the small metal tab. The float moves up or down in the float bowl according to the level of gasoline. At a certain level, the needle valve pushes into its seat tightly enough to shut off any further flow of gasoline. As the engine consumes fuel, the level of fresh fuel in the bowl diminishes. The float pulls the needle away from its seat, allowing more gasoline flow into the bowl.

To remove the float from the carburetor, simply pull out the pivot pin with a pair of long-nose pliers, as shown in Fig. 25. Be careful that the float does not fall suddenly and bend or break any part of the assembly.

After the float is off, slip the needle valve gently from its seat. Look at the taper on the end of it to see if a groove is worn on the taper where it touches the seat. See Fig. 26. If the taper has a groove, the valve can leak even when the needle is pushed against the seat. This wastes gasoline by overfilling the float bowl.

To replace the float and needle, first put the float back in place and push the pivot pin through its mounting holes. Then carefully slide the needle back into its hole in the carburetor (See Fig. 27.) This is easier if you lift the float back out of the way with one hand and guide the needle with small-nose pliers in the other hand. When the needle is partly in place, slip the small retaining tab over the end of the float tang. Then release the float, and all the parts should pop into place at the same time. Work the float up and down a few times to make sure it and the needle valve move freely.

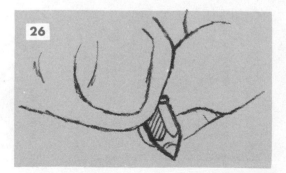

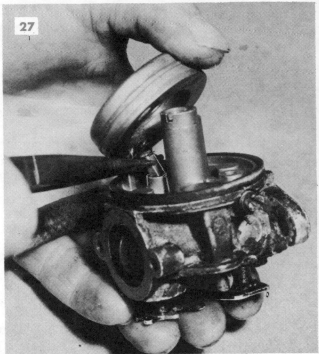

27

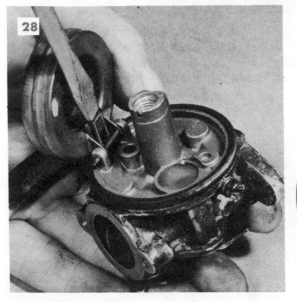

28

24

The only way to adjust the float is by bending the tang the needle valve rests against. Place the end of your screwdriver under or against the tang and press just hard enough to move the tang a tiny amount. See Fig. 28. This changes the level of gasoline the float allows in the bowl. Do not bend the tang much; a small change at the tang produces a large change in gasoline level in the float chamber.

Before you reinstall the carburetor on the engine, manipulate the throttle arm on the outside of the carburetor a few times to make sure it moves freely. See Fig. 29. Look inside the throat of the carburetor at the throttle butterfly. It must turn freely, with no obstructions to block its free movement. Its mounting screws should be tight.

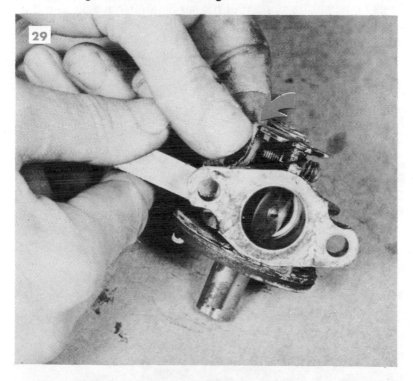

Now bolt the carburetor back onto the engine. Hook everything up the way it was: kill wire, governor link, throttle cable if there is one. Open the gasoline shutoff valve. Wait a few seconds for the float bowl to fill with gasoline. Then push the float bowl drain valve (Fig. 30) to make sure a plentiful supply of gasoline reaches the carburetor through the fuel line from the gas tank.

THE SIPHON-FEED CARBURETOR

Working on a siphon-feed carburetor demands the removal of the air cleaner. It is held, usually, by one large-head bolt through the center of the cover, as shown in Fig. 31. A large rubber gasket fits between the carburetor and the air cleaner, as shown in Fig. 32. The gasket damps vibration, and keeps the air cleaner bolt from shaking loose. It also seals between the bottom of the air cleaner housing and the carburetor. Thus no air can enter the carburetor that has not passed through the filter.

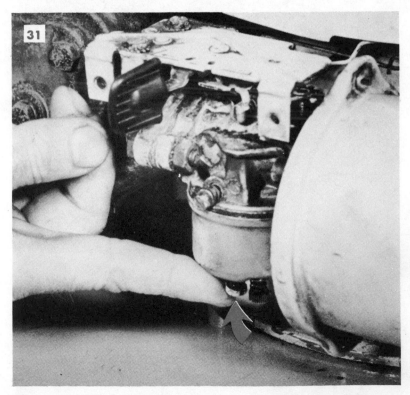

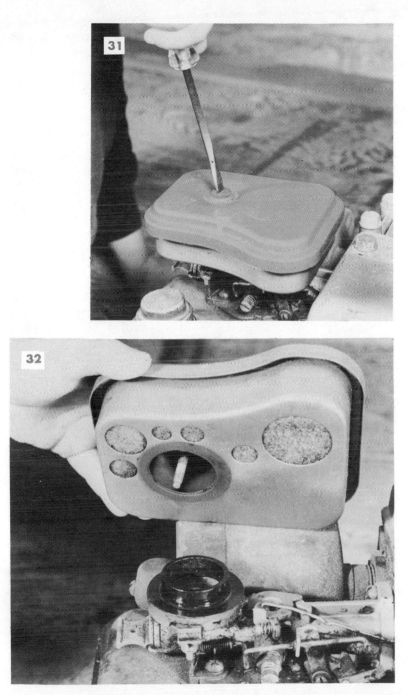

Unfiltered air entering the carburetor carries minute particles of dirt. These particles induce excessive wear in the carburetor and engine. Never operate any engine without the air cleaner on—except when you're working on the engine, and then only as much as is necessary.

With the air cleaner removed from the carburetor, separate the top and bottom sections to expose the filter, as shown in Fig. 33. The filter itself is a piece of foam rubber soaked in oil. The oil traps dust particles from the air that passes through. The filter should be washed in gasoline to remove dirt and grime and then resoaked in oil. A steel spacer tube through the center of the foam rubber, shown in Fig. 33, keeps the engine vacuum from sucking foam rubber pieces down into the carburetor. Wipe this tube clean every time you remove the filter for servicing.

You can now remove the fuel tank from the carburetor. The two screws (Fig. 34) that hold the tank are accessible only after the air cleaner has been removed. A gasket between the tank and the bottom of the carburetor form an air- and dust-tight seal. You do not want dirt in the gas tank.

After the gas tank is out of the way, you can now remove the carburetor from the engine. The bottom screw is in such

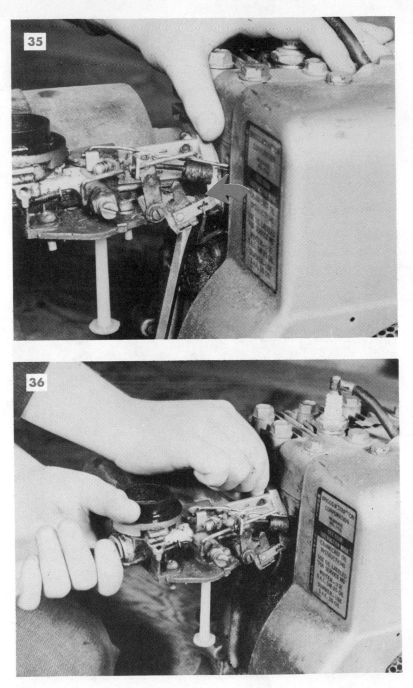

a tight place and should be removed first as shown in Fig. 35. You extract the top screw by hand after you loosen it with the screwdriver. Always hold the carburetor with one hand as you take out that last screw, so the carburetor will not fall and be damaged. See Fig. 36.

A small rod hooked to the throttle connects it to the governor. See Fig. 37. This rod has a small kink in the end, to hold it in the throttle lever yet lets it maneuver freely. You must tilt the carburetor sideways to slip the rod end out of the mounting hole. Be careful and don't bend the rod as you disengage it. That would change the operation of the governor.

Once the carburetor is off, you will marvel at its simplicity: a barrel with a plunger to let fuel-air vapor into the engine. Beyond what has been described, analysis follows that for a gravity-feed carburetor.

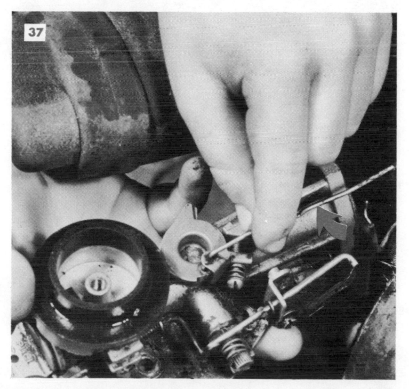

A siphon-feed carburetor has only one mixture screw. It controls the air-fuel ratio both at idling and at operating speed. Like the mixture adjustments detailed earlier, this one should be altered only when the engine is running. When the screw (a needle valve) is set properly, the engine idles smoothly, accelerates evenly when you open the throttle, and runs steadily at operating speeds. See Fig. 38.

If you suspect a bad needle valve, count the number of turns (clockwise) it takes to just touch the seat. Then back it out and inspect it as described earlier for gravity-feed carburetors.

An idle-speed screw is the only other adjustment on this carburetor. It forms a stop for the throttle linkage. Set it so the engine idles as slowly as it can without trying to die or run rough. Try to correlate mixture and idle speed for smoothest idling commensurate with positive acceleration.

Cleaning inside and out constitutes the major repair you can apply to this type of carburetor. Soak it an hour or two in carburetor cleaner and then wash it off with clean gasoline. That removes all the dirt and sediment from inside and takes off all the scum deposited inside or out. Be sure the screen at the bottom of the siphon tube is absolutely clean; otherwise the carburetor cannot lift enough gasoline to let the engine run.

TWO-CYCLE CARBURETOR

The carburetor for a two-cycle engine comes close in design to that for a four-cycle engine. The major difference lies in all of the internal holes and passages. These holes are larger, to allow for the heavier fuel, which forms larger, thicker droplets. Oil must be mixed with the gasoline in a two-cycle engine. All two-cycle carburetors are gravity-feed. It would be very hard, during starting, for the engine to lift the heavy gasoline-with-oil mixture.

As usual, the first step toward repair of the carburetor is removal of the air filter. The top half of the air-cleaner housing, illustrated in Fig. 39, is held in place by a pair of spring clips. When you release the top of the housing, you can take out the foam rubber filter. Then remove the two

screws that hold the cleaner housing to the carburetor. See Fig. 40. This lets you set the housing out of your way while you work on the carburetor.

Remove the float bowl from the bottom of the carburetor. The bolts that hold the bowl in place on this particular engine have an L-shaped extension instead of a screw-slot head—you can remove them with a pair of pliers. See Fig. 41.

After removing the float bowl, check the float needle valve just as you would with the four-cycle carburetor. Be sure both float and needle move freely and are not worn, bent, or scored. You may find a float made of cork, instead of a hollow metal "bulb" or donut. Check the cork for oversoaking. It can be so heavily oil-soaked that it will not float on top of the gasoline. In that event, you can clean it by soaking it overnight in alcohol or lighter fluid. That usually removes the oil.

Sometimes, the cork flakes off in small pieces. The particles clog up the carburetor and no fuel or vapor can get through. The only cure is a new float. See Fig. 42.

You can remove and clean the main jet in a two-cycle carburetor. It controls how much fuel is injected into the carburetor barrel when the engine is running. A screwhead in the center of the float ring gives you access. Just back it out (counterclockwise) and check it every time you take the

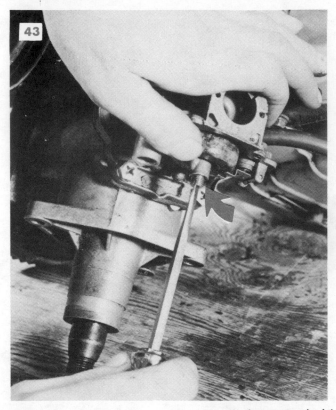

carburetor apart for any reason. You'll have to hold the float up to make the head of the screw visible. Be careful not to damage the slot in the end. The jet is made of brass, a soft metal that can be damaged easily. See Fig. 43.

Inspect the jet first for wear or damage. See Fig. 44. A common carburetor problem in two-cycle engines stems from this jet becoming plugged with sludge. Oil in the gasoline mixture absorbs moisture whenever you store the engine for a period of time, and the glop thus formed builds up in any tight passage it can find.

If a two-cycle engine seems hard to start, yet produces a good spark, this jet probably is stopped up. Pull it out and clean it thoroughly with fresh gasoline. Then blow through it with your mouth or with compressed air if any is available (Fig. 45). The forced air should remove any tiny particles of dirt or sediment trapped in the orifices.

36

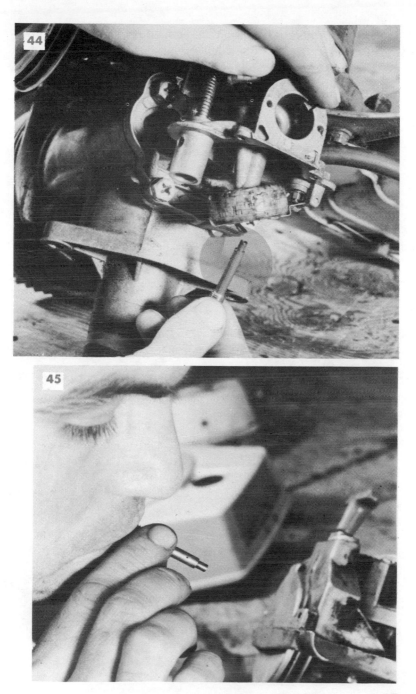

The run-mixture adjustment on the carburetor shown in Fig. 46 has a large black plastic knob instead of a screwdriver slot. As always, do not mess with the mixture screws unless the engine is running. Only turn the mixture screw a little at a time and give the engine a moment to adjust.

The adjustment with the white knob is for the idle mixture. See Fig. 47. You set it with the engine running at idle speed. Adjust both mixture screws alternately, until the engine runs smoothly at all speeds.

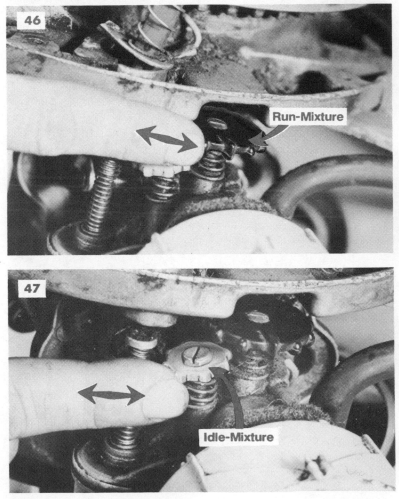

There is one other control knob on this carburetor: the *primer knob*. When you start a cold engine, you depress the primer knob a few times (Fig. 48) to force some raw gasoline into the piston compartment. The incoming air, drawn by cranking the engine, carries some of the vapor directly into the cylinder above the piston. This bypasses the carburetor, giving the cylinder a more volatile charge for easier starting. This knob also works the kill switch for the engine. When you rotate the primer knob, a metal tab on it touches the kill wire from the coil. This tab shorts out the magneto coil when it touches the wire.

48

Primer Knob
And Kill Switch

Any foam-rubber air filter should be cleaned thoroughly by soaking it in gasoline at least once every 25 hours of operation. Pour some clean gasoline into a coffee can. Immerse the filter, then squeeze the gasoline out of the filter. Repeat the soaking and squeezing until the filter comes clean. See Fig. 49.

Afterward, pour new motor oil all over it, soaking it thoroughly (Fig. 50). The oil in the foam rubber is what catches the dust particles from the air that enters the carburetor. The oil holds these dust particles in the filter. A filter full of dirt severely reduces the amount of air that can pass through to the carburetor. The carburetor cannot deliver a properly lean fuel-air mixture and the engine runs poorly. What air does get through is still dirty. Carburetor and engine both suffer.

Tune-ups and Ignition Repairs

FOUR-CYCLE—VERTICAL-SHAFT

The first step in diagnosing any ignition trouble is to check the condition of the ignition spark delivered to the engine's combustion chamber. The clip that connects the high tension wire to the spark plug is a good place to start. It should be tight on the top of the spark plug.

Test the spark. Detach the spark plug wire and hold it approximately ⅛ inch from the metal cap of the spark plug. Take care that your fingers do not touch the metal clip, or you may get a sharp jolt. Pull the starter rope a few times. A bright, hard, blue spark should snap across the gap between the wire and the spark-plug cap. A weak spark exhibits a yellow color. It leaves the engine hard or impossible to start. See Fig. 51.

If the spark is in good condition, continue the sequence. Check the spark plug itself. Remove the spark plug from the engine. A special spark-plug socket wrench is handy to remove the plug from a small engine, but it is not absolutely necessary. Be careful removing the spark plug, and do not break the porcelain. A spark-plug socket with a rubber insert inside, a worthwhile purchase, can help prevent breakage. See Fig. 52.

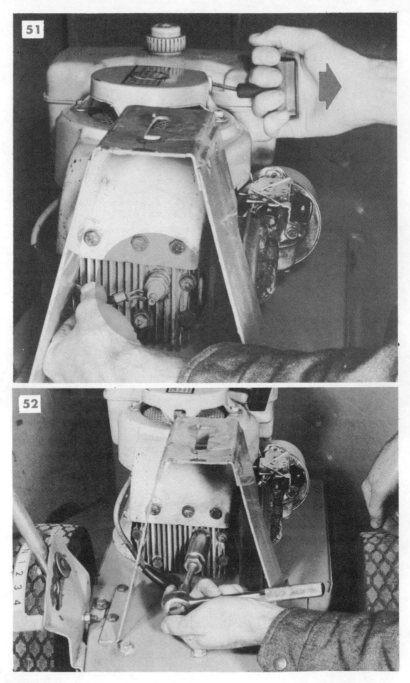

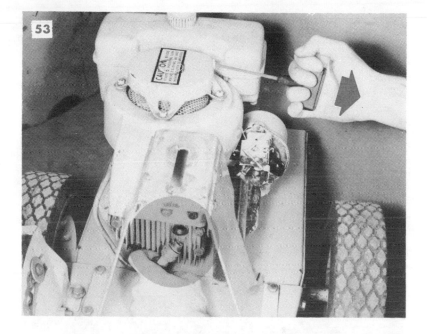

After you have removed the spark plug, reattach the heavy wire. Then hold the spark plug against the engine with the threaded steel base touching the engine. Press it firmly, to provide a good electrical connection. Watch the bottom tip of the spark plug as you pull the starter rope a few times. You should see the same bright blue spark jump across the gap between electrodes at the tip of the spark plug. A strong spark makes an audible snapping sound. The spark plug is in good condition if the blue spark is as snappy as it was at the tip of the plug. If you see a yellow spark, or no spark at all, replace the spark plug. Always buy the type (number) of spark plug recommended by the manufacturer of the engine. See Fig. 53.

While the spark plug is out, you should test compression. Here's a fast and acceptable method of checking the compression. Hold one finger over the spark-plug hole and pull the starter rope. The piston should develop enough compression to blow your finger off the spark-plug hole. If you do not find much compression, take the engine to a reputable repairman. He will install new piston rings and grind or

45

replace the valves. Worn rings or burned valves cause poor compression of the air-fuel mixture, which makes the engine hard to start. See Fig. 54.

After you complete the spark and compression tests, remove the starter to give you access to the breaker points. You should replace the points at least once a year—generally, at the beginning of the season—even if the engine starts and runs. New points and a new spark plug save gallons of gasoline in a season, and make the engine easier to start.

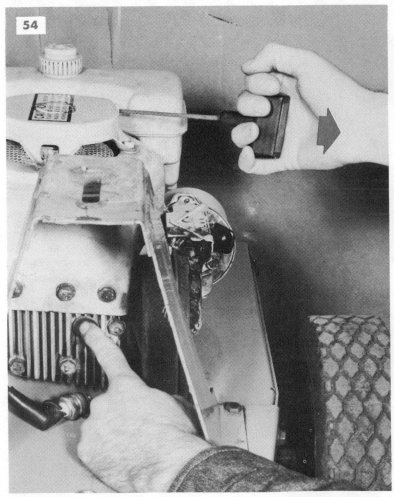

54

The starter usually mounts on the flywheel cover. On some engines, the starter housing has been molded as part of the flywheel cover instead of bolting on separately. See Fig. 55.

On the engine used here as an example, one end of the flywheel cover bolts to the crankcase. The other end is held in place by the cylinder-head bolts. Be careful when you replace the head bolts, after working on the engine beneath the flywheel cover, that you tighten them properly. They should be tightened only to a wrench torque of 20 or 25 foot-pounds. Too much torque warps the head; not enough causes the head gasket to blow out soon. A special torque-wrench is best for this. See Figs. 56, 57, and 58.

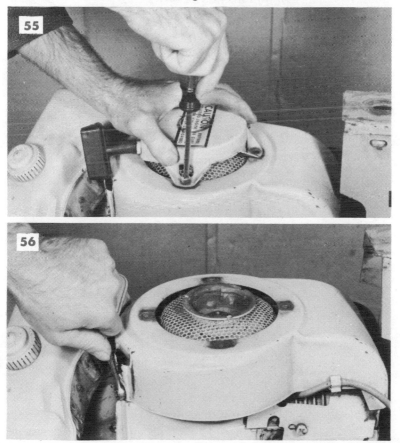

With some riding mowers, you will have to loosen the engine mounting bolts and move the engine enough to let the flywheel cover slide off. Be careful as you lift the cover. Do not bend any part under it. The governor may be mounted underneath, right beside the flywheel. An airflow governor can be bent all to easily.

There are many variations in the shape of flywheel covers and where the bolts are placed. But the basic design is pretty much as shown here. Also, on this particular model, the gasoline tank must be removed. On some models the gasoline tank is mounted separately from the engine.

At this point, before any other work is performed, it is a good idea to clean the air fins of any accumulated grass and debris that has been trapped under the flywheel housing. See Fig. 58.

Next, with the flywheel cover off, you are ready to loosen the nut that holds the starter catch, the leaf screen, and the flywheel on the end of the crankshaft. You will have to find a way to hold the flywheel steady while you break loose the nut.

Try placing the end of a wrench between two of the cooling fins of the flywheel. If you do not have a wrench that can fit between the fins without slipping, try a large screwdriver or a bar. Anything solid that will not slip should be sufficient.

Back the nut off (counterclockwise). Then lift the starter catch-plate and the leaf screen off. Set them aside until you are ready to put everything back together. Be very careful not to break any of the flywheel cooling fins. See Figs. 59 and 60.

A tapered hole in the center of the flywheel matches the tapered end of the crankshaft. A keyway provides positive alignment for the flywheel, and the taper helps hold the flywheel nut, the taper fits so snugly together, removing the flywheel takes a vigorous tap to shake the two surfaces apart.

To accomplish this safely and easily, screw the nut back on the crankshaft until the end of the crankshaft is exactly flush with the face of the nut. Insert a screwdriver under the flywheel and pry upward just enough to put pressure

48

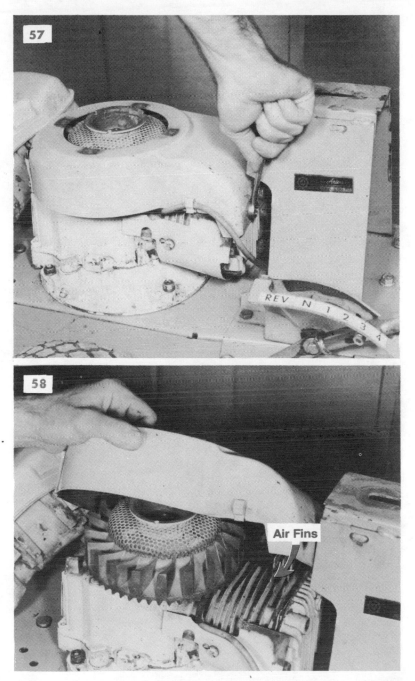

Air Fins

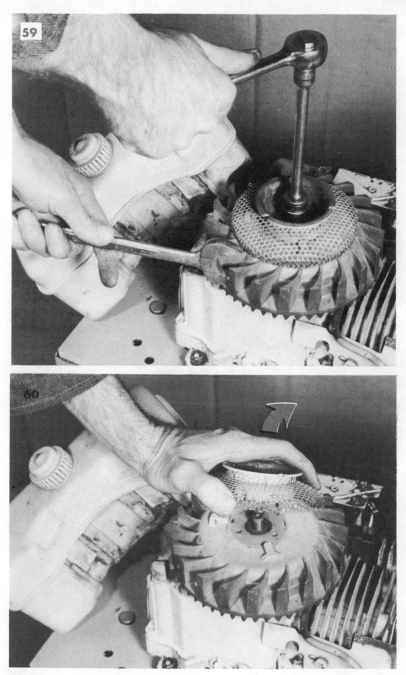

on the flywheel. Then hit the end of the crankshaft with a hammer—not a crushing blow, but a sharp rap. The tapers usually separate immediately. Be sure the nut and the end of the crankshaft are flush, or the threads on the crankshaft may be damaged by the hammer. See Fig. 61. Once you have jarred the flywheel loose from the taper on the crankshaft, it can be lifted up and removed. See Fig. 62.

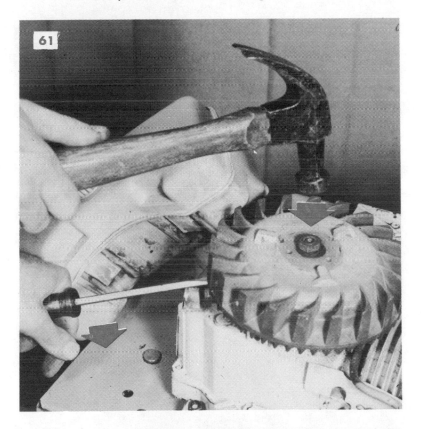

As you lift the flywheel clear of the crankshaft, be sure the tiny metal key that aligns the flywheel on the crankshaft does not fall out and get misplaced. If for any reason you have to replace this key, be sure to purchase a new one from a dealer. The key is made of nonmagnetic metal. A steel key would interrupt the magnetic field for the magneto, and the engine might run poorly or not at all.

Turn the flywheel over and inspect the magnets mounted inside. These magnets must be clean and free of any large nicks. They are part of the magneto that generates the ignition spark to run the engine. See Fig 63.

The next part you take off is the breaker points cover. It is held in place by a spring clip, which you can snap free by prying up one side with a screwdriver. Turn it around to clear the cover. See Fig. 64.

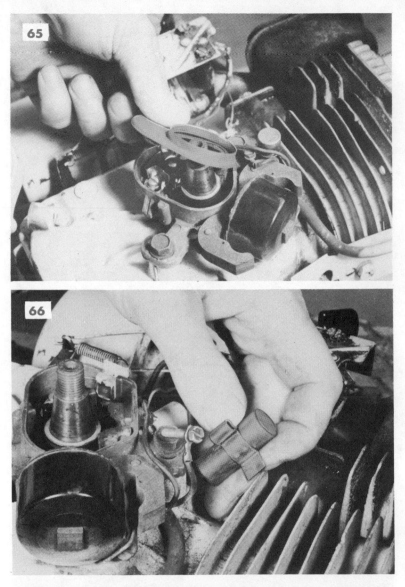

Lift the cover off and lay it aside. See Fig. 65. Be careful not to tear or bend the gasket you find under the cover. This gasket seals the breaker point compartment. It keeps out dust and small trash that could short out the points, keep them from closing properly, or wear them unduly.

54

The condenser is the next item you remove. It is situated just outside the bracket that holds the points. One screw holds the condenser in place. This same screw holds the point ground wire. Be sure you remember which wire goes under this screw. Draw a diagram if you need to. See Fig. 66. After you take out the screw, pull the condenser and ground wire away from the engine slightly to leave you room to work.

A defective condenser usually causes the contact area on the breaker points to burn and become pitted rapidly as the engine runs. A weak spark, one that appears yellow at the spark plug, is often the fault of a poor condenser. Replace the condenser every time you install points.

The nut that attaches the condenser wire to the breaker-points terminal also holds a wire from the coil and one from the kill switch (on the throttle controls). This nut must be loosened for you to replace the points and condenser. A small ignition wrench fits this nut. (A set of ignition wrenches contains several sizes to fit these small nuts. A set is not expensive.)

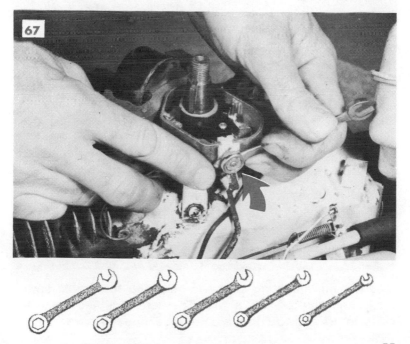

Remove all wires from the bolt and fold them gently back out of your way. Do not bend them sharply or very often, you can break them off or damage the insulation. Observe and remember which wire fits under this nut, and in what order, so you can replace them the same way. See Figs. 67 and 68.

Remove the breaker-points mounting screw. Later, this is the screw you will leave just a little bit loose when you set the points gap. Be careful; it's easy to drop this screw down the crankshaft hole into the engine.

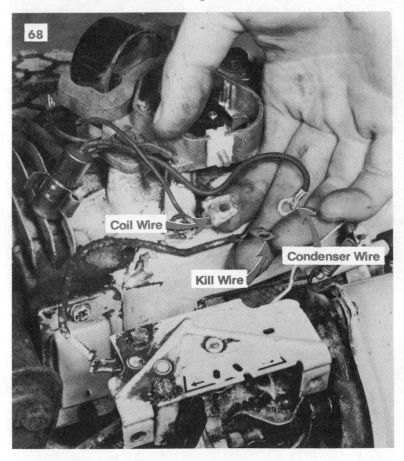

68

Coil Wire

Condenser Wire

Kill Wire

After the mounting screw is out, take the points spring loose from the terminal screw. They are together inside the housing to provide the electrical connection from those

56

external wires to the points contacts. You can lift up the spring with the tip of a screwdriver. See Figs. 69, 70 and 71.

Lift out the points. Inspect them thoroughly. If the contact faces are burned or glazed, or if they appear pitted or uneven across the faces, replace the set.

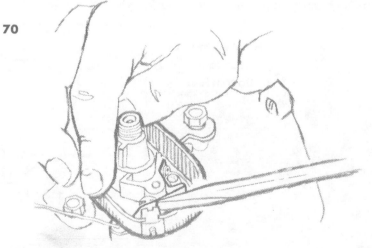

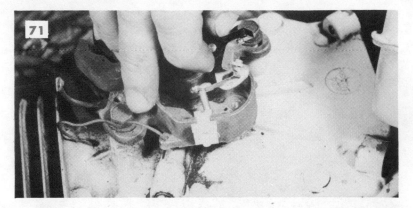

Breaker points for small engines are usually in two separate pieces. The top half swivels on a small post that is part of the bottom half. See Fig. 72. The fixed nonmovable bottom half of the breaker points set is made of steel so it stays rigid and grounded. It contains one contact face. The other half of the points is made mostly of plastic. The plastic body rides on a cam that is part of the crankshaft. Firmly a part of the plastic body is the other contact face and the metal spring that goes to the terminal bolt. The spring, besides making the electrical connection, holds the plastic body firmly against the eccentric cam on the crankshaft. Fig. 73 shows the cam which opens and closes the breaker points. This controls the shape and duration of the magneto pulse that creates the ignition spark.

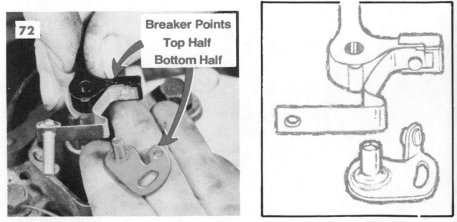

Breaker Points
Top Half
Bottom Half

Before you can repair or replace a faulty magneto coil —an uncommon occurrence—you must take the breaker cam off the crankshaft. The cam is a ring that slides onto the crankshaft. You can see or feel the lump or high place on the outside that moves the breaker-point contacts apart.

A key aligns this ring on the crankshaft. See Fig. 73. It is the same key that positions the flywheel on the crankshaft. The relationship thus established between cam ring and flywheel "times" the ignition spark. This timing is fixed and cannot be altered by you or a repairman as it can in an automobile engine.

You must pull the key out before you can slide the cam ring off of the crankshaft. Be careful not to drop or damage the key. It cannot be damaged in any way, because it fits very snug in the flywheel and crankshaft.

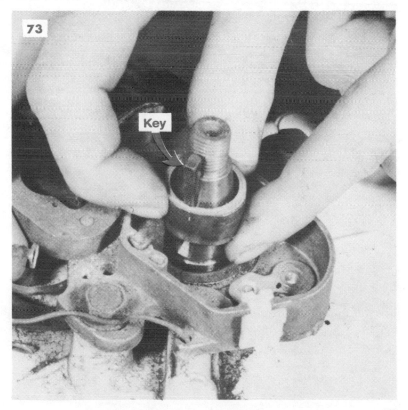

Remove the two bolts that hold the coil mount to the engine and remove the coil. The coil mount in this instance is part of the breaker-points housing. See Fig. 74.

If you remove the coil, do not disturb the marks left on the mounting bracket by the heads of the bolts. They can guide you in replacing it. See Fig. 75.

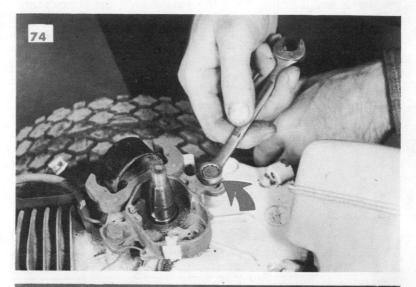

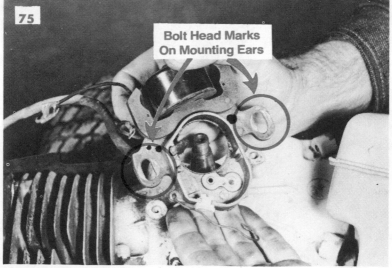

Bolt Head Marks
On Mounting Ears

If you have any doubts about the condition of the coil, take it to a small-engine repairman, he can test it for you. Otherwise, install a new coil—an exact replacement bought from a dealer. Make a careful note of how the wires are situated around the coil bracket, so you can replace them in the same position.

After either the old coil or a new one has been installed, you must set the gap between the magnets in the flywheel and the magnetic poles on the coil. Snug up the mounting bolts of the coil, but leave them so you can move it while the flywheel is down over it. Place the flywheel on the crankshaft.

Turn the flywheel (and crankshaft) so the magnets are adjacent to the coil poles. You adjust the gap between poles and magnets by moving the coil physically. You need a feeler gauge or something to "feel" the gap. You must set the gap between the coil and flywheel magnets to .020 inch. See Fig. 76.

76

Use the .010 blade in a steel-leaf feeler gauge, since curvature of the flywheel would make the .020 blade push the gap too wide. If you have no feeler gauge, a cardboard matchcover has almost the right thickness. Set the coil for a loose "feel" if you use a matchcover.

Now lift the flywheel off the crankshaft again. If you are sure the coil gap has been properly set, tighten the mounting bolts securely. Install the cam ring and the key that positions it on the crankshaft.

Now you are ready to install the new set of breaker points. Replace the assembly just as it was removed, except in reverse. Be sure the spring hooks over the terminal bolt. With the points in place, turn the engine crankshaft until the high surface on the cam ring presses on the plastic body and separates the point contacts. An open-end wrench ($\frac{9}{16}$") works fairly well to turn the crankshaft. This method will not damage the crankshaft in any way, but be careful not to distort the key. The crankshaft turns very easily if you take the spark plug out temporarily so there is no compression. See Fig. 77.

When you have turned the crankshaft so the cam ring opens the points, you can set the gap between the contacts. Look for a small notch at the base of the points. A screwdriver fits that notch and lets you slide the baseplate of the points assembly to the proper setting.

Insert a clean feeler gauge between the contacts. The gap should be set to .020 inch for most small engines. See Fig. 78. Let the contacts just barely touch the feeler leaf. Be careful that the gap remains correct as you tighten the screw that locks the base. Always turn the crankshaft a few times and recheck the widest position of the gap, just to be sure it is correct.

Next, install a new condenser. Be sure the coil ground wire connects to the proper place. Check that no wires are pinched. The coil wire in this engine lays right beside the condenser, and the condenser holds the wire in place.

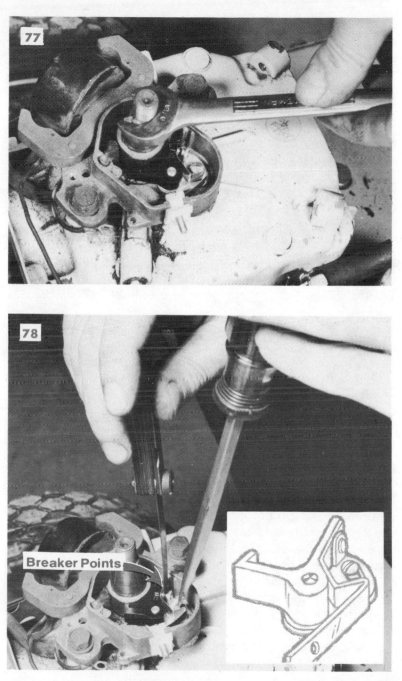

Breaker Points

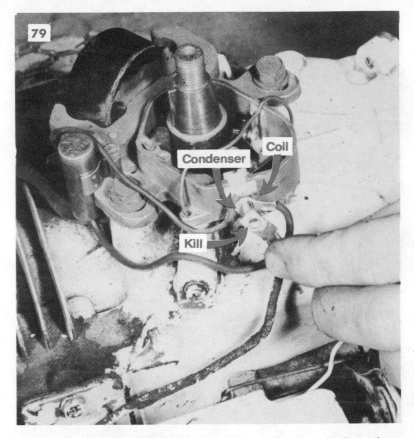

The next step in reassembly is to replace all the wires on the breaker-points terminal bolt. There should be three wires on this terminal: the coil wire, the condenser wire, and the kill wire. See Fig. 79. Position all of them so they do not bind anywhere. Do not let their lugs touch any metal except the terminal bolt—and of course the nut.

Replace the nut and tighten it. Make sure the wire terminals do not turn and pull any of the wires into a bind as you tighten the nut.

Now seat the cover over the points compartment. The fishpaper gasket that fits under this cover must not be broken or bent and must fit into place properly. It seals out dust and debris. It also insulates against the possibility of a spark jumping from the points to the metal cover. See Fig. 80.

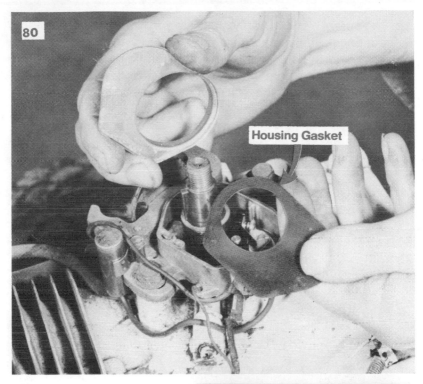

Housing Gasket

Slip the remaining spring clip back into place, as shown in Fig. 81. It fits into a groove and presses tightly against the cover. It must hold the cover securely so vibrations of the engine cannot shake it loose.

You are now up to the last step in your points changing sequence. Replace the flywheel. See Fig. 82. Set it carefully down over the crankshaft. See that the key fits smoothly into the keyway in both crankshaft and flywheel. Replace the grass screen and the starter catch on top of the flywheel. See Fig. 83.

Screw the flywheel nut on as far as you can by hand. To tighten the nut well, use the same open-end wrench you held the flywheel with to remove the nut. Replace the flywheel cover and you are finished. Do not forget to use a torque wrench on the cylinder head bolts—20 to 25 foot-pounds.

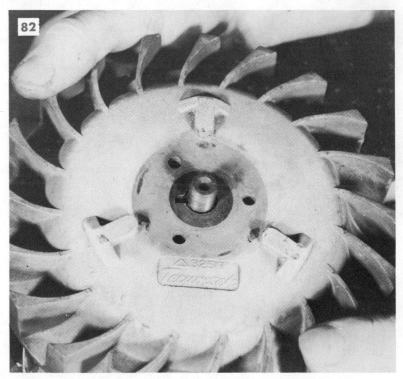

82

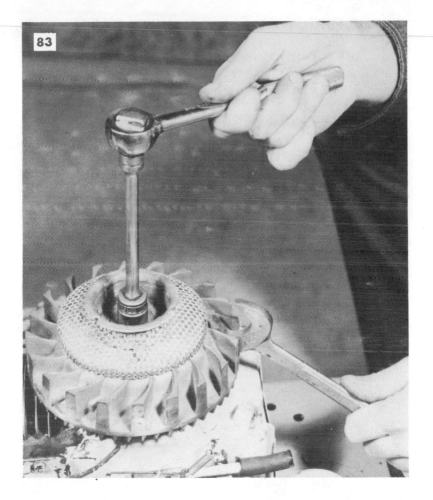

FOUR-CYCLE—HORIZONTAL SHAFT

Let us look at tune-up and ignition repairs in another type of small engine—a horizontal-shaft four-cycle. You will see some minor variations in starter design and in the shape of the breaker points.

Almost without exception, you start any ignition system troubleshooting by removing the spark-plug wire from the top of the spark plug, as shown in Fig. 84. Hold the wire clip about ⅛ inch away from the end of the spark plug. Pull the starter a few times. If you can see a bright blue spark, then the points and magneto are working properly. You should also be able to hear a snapping sound when the spark jumps the gap between wire and plug. If the spark looks weak and yellow, or if you see no spark at all, something is wrong with the points, condenser, or coil.

84

A visual inspection at the bottom end of the spark plug usually tells you if you need a new plug. It can also tell you why the plug became defective. If the tip looks black and parched, the carburetor may be set too rich. The engine does not burn all the fuel delivered to the combustion chamber. Correct that failure after you have finished your ignition checks. If the electrode tips are burned down close to the porcelain, the spark plug is simply worn out. Install a new plug.

If the electrodes look okay, verify that the plug can handle a good spark. Attach the wire with the spark plug out of the engine. Hold the threads against the engine metal. Pull the starter a few times. If the bright blue spark jumps across the electrode gap, and makes a clear snapping sound, the spark plug works fine. If the spark is yellow or missing, yet was okay at the plug wire, you need a new spark plug. Always buy whatever spark plug the manufacturer recommends for your small engine.

Before you install a new spark plug, set the gap between the center and ground electrodes. The proper gap for single-cylinder four-cycle engines is .030 inch. Some repairmen gap the spark plug to .025 inch, but the engine starts easier and runs more efficiently with the gap at .030. Any gap between .025 and .030 suffices to start the engine and make it run. See Figs. 85 and 86.

85

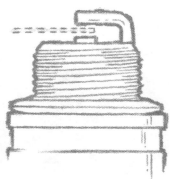

Use a spark-plug feeler gauge for gapping a spark plug. It is an inexpensive tool, and gives you a true reading in the gap. This tool usually has a notch blade to bend the ground electrode with. This is how you alter the gap—never hit the ground electrode on a solid object to change gap. Some spark plug feelers have wires instead of blades to measure the gap. Either type does the job well.

Before you install a new spark plug, ascertain that the washer is there and rotates freely above the threads. See Fig. 87. When you screw the spark plug into its hole in the head, start it by hand so it cannot get cross-threaded. That could ruin the cylinder head or the spark plug or both.

Use a deep spark-plug socket, preferably with a ratchet handle or a torque handle, to tighten the spark plug. If you use a plain wrench, it can so easily slip and crack the porcelain part of the spark plug. If the porcelain is broken, the plug is ruined. Even worse is a porcelain crack you cannot see, yet it keeps the engine from starting (maybe only on damp mornings).

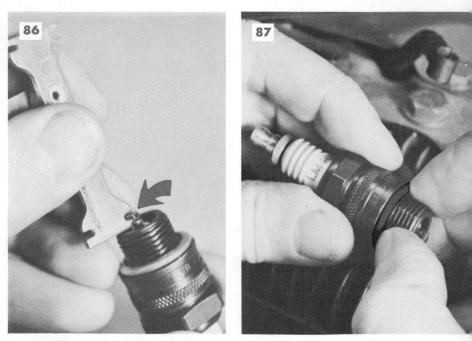

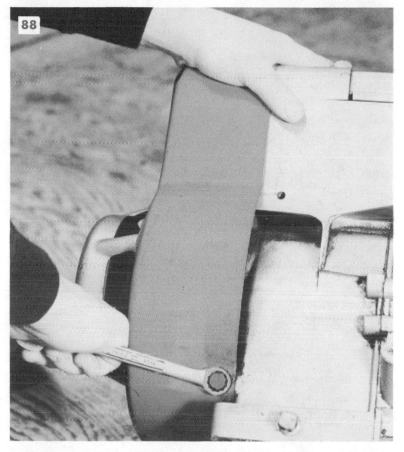

After you have checked and/or replaced the spark plug, direct your attention to the breaker points and condenser. You must remove the flywheel cover first. On this particular engine, the starter housing and the flywheel cover form one unit.

Remove the bolts that hold the flywheel cover to the engine. Two are on the sides of the cover, down low. The others are on top. None of the bolts that hold this cover are head bolts, so you have no worry later in retightening them to a specified torque. Be careful as you remove the cover. Do not bend the governor controls. The governor parts are directly under the flywheel cover, and susceptible to damage. See Figs. 88 and 89.

Before you set the flywheel cover aside, take a look at how the starter is built. It works with a ratchet mounted on the end of the crankshaft. A square spindle protruding from the ratchet fits into a square opening in the starter pull assembly. See Fig. 90.

The next item to come off usually is the leaf screen that covers the top of the flywheel. Four small screws hold it onto the flywheel (instead of the flywheel retaining nut holding it as you saw on the vertical-shaft engine). See Fig. 91. This screen keeps trash from working in next to the fins on the flywheel. That could impede air movement around the engine and cause overheating. An overheated engine is quickly ruined.

Having the coil mounted outside the flywheel is another variation in design. This permits you to remove the coil for testing or replacement without having to remove the flywheel first.

The coil mounting bolts are easy to reach. See Fig. 92. Be sure to note whether the ground wire is held by one of the coil mounting bolts—usually it is as shown in Fig. 93. This ground wire provides a solid electrical ground for the coil. If it is not securely under that bolt, the engine cannot run.

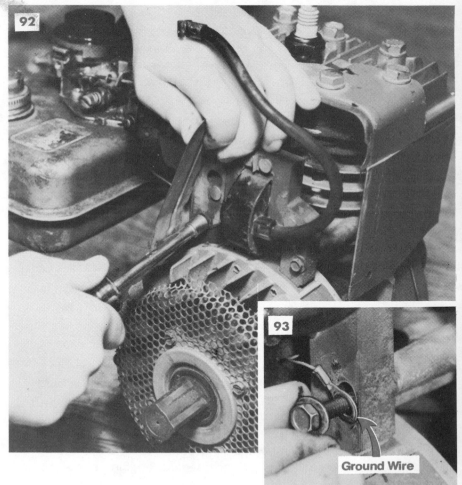

92

93

Ground Wire

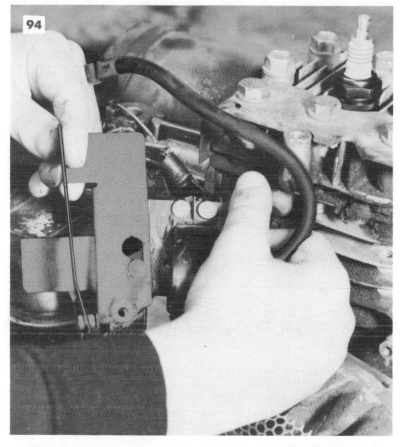

The governor plate mounts with the same bolts that fasten the coil to the engine. Be very careful handling the governor, as it can be bent very easy. See Fig. 94.

The magnets that are mounted in the flywheel are one-half of the magneto. You can check them by holding a screwdriver or other steel tool near them. Feel how strong their magnetic pull is. It should be strong enough that you have to exert some force to pull the tool away.

The coil is the other half of the magneto. It must be mounted so its pickup faces are .020 inch from the outside rim of the flywheel. Two strips of manila file folder make a convenient feeler gauge if no other is available, as shown in Fig. 95.

On this model engine, the ratchet assembly for the starter serves also as retaining nut for the flywheel. There is no hexagon nut on the ratchet, so you cannot remove it with a wrench. The square part only "catches" in one direction, since it is a ratchet.

On many models, this ratchet is locked in place by a large washer with square teeth installed between ratchet and flywheel. When ratchet is tightened sufficiently, one or more square teeth are bent up into a corresponding notch in the ratchet; thereby locking the ratchet in place.

On this model a screwdriver is placed on one of the ears on the outside of the ratchet housing, as illustrated in Fig. 96. Extreme care should be used when hitting the screwdriver—the ratchet ears can be broken very easily. When the ratchet housing loosens, you can unscrew it from the crankshaft. Remove the whole ratchet assembly. See Fig. 97.

Use your hammer-and-screwdriver (as explained previously) method to pull the flywheel. However, there are no threads on the end of the crankshaft to be damaged. But be careful hitting the end of the crankshaft; do not mar it. Hit the crankshaft with short, sharp raps—not heavy, crushing blows. It would be even better if you use a rubber or plastic mallet, or hold a block of wood against the crankshaft.

When you have loosened the flywheel, slide it off of the crankshaft and put it aside. See Fig. 98. The points cover is removed next. It has been mounted under the flywheel to protect it and the breaker-points assembly from damage.

Pause in your disassembly long enough to clean up the engine. Wrap a cloth around a screwdriver blade and wipe out all the small hard-to-reach places on the engine. Try to remove all the accumulated grease and dirt. There are two reasons to do this. First, a clean engine is much easier to work on than a dirty one. Second, the dirt keeps engine heat from being carried away and dissipated. This is very important in keeping the engine from wearing out prematurely.

Hold your hand below the points cover to catch the screws that hold it on. They will probably fall out as you remove the cover. See Fig. 99.

There is one design factor very different in this engine. The condenser is of special design. One of its terminals, the one that otherwise would be connected by a wire, forms one contact for the breaker points. So, in tuning up for the season, the condenser is the first portion of the points assembly you remove. The condenser is held in place by a clamp with one screw. See Fig. 100.

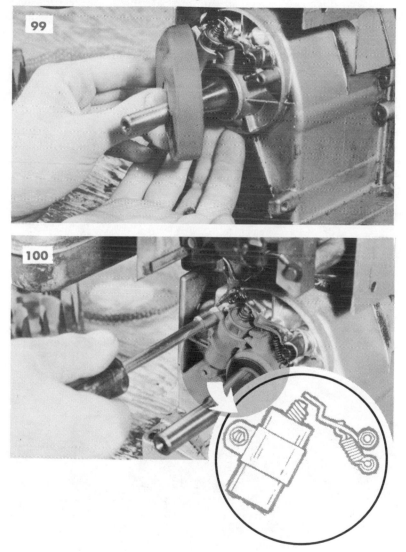

You can see the unusual design of this condenser, compared to those shown previously. Also, the wire from the coil attaches to the condenser instead of at a breaker assembly terminal. The coil wire is held in place by a small compression spring clipped over the stem of the condenser, compressed, and then allowed to expand (relax) to hold the wire in a small hole in the contact stem. See Fig. 101.

Actually, two wires terminate at the point/condenser stem: the coil wire and the kill wire. To get the two wires loose from the condenser, hold the small spring down with the end of a screwdriver or other flat tool and pull the wires out of the hole. Catch the spring. With no wires to hold it compressed on the stem, it flies off. See Fig. 102.

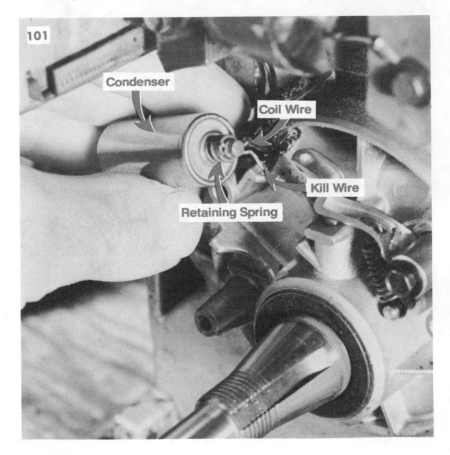

The breaker arm is the other half of the points contact pair. It is held in place around a ground post by a small but heavy expansion spring. See Fig. 103. This breaker point assembly resembles very little the one shown previously. A pair of sharp-nose pliers seems like

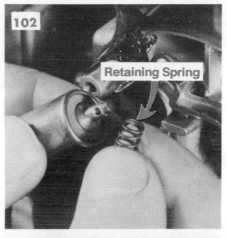

the best tool for removing the breaker arm from its ground post. See Fig. 104.

Remove the breaker arm spring next. Pull the spring off the pin it hooks over. Then you can take the bolt out of the ground post itself, and release the ground wire. One end of this wire is attached to the breaker arm; the other is held against the engine frame by the bolt through the ground post. See Fig. 105.

Be careful not to break this ground wire. It is relatively thin, and has no insulation, so it's a bit fragile. This wire must make solid connections at both ends, since it is the chief ground for the breaker arm. See Fig. 106.

The ground post has a groove along one side. The end of the breaker arm rests in this groove, which forms the swivel point. The breaker arm pivots there when the cam moves it. A small key-like protrusion cast in the mounting holds the ground post in one fixed position. This "key" catches the end of the groove when the ground post is bolted solid.

A small rod between the breaker arm and the crankshaft cam pushes the movable breaker contact away from the fixed condenser contact, each time the high spot on the cam comes around.

To install a new breaker arm, first slip the ground wire under the ground post and firmly tighten the post bolt. Then place the end of the new breaker arm in the groove of the ground post. Hold it down to the bottom of the groove. Pull the heavy spring down and slide its loop over the pin. You may have to use sharp-nose pliers to stretch the spring enough to slide over the pin. See Fig. 107.

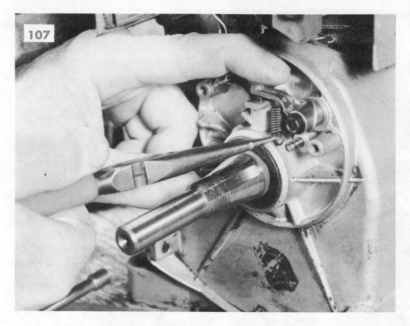

Many servicemen still file contact faces on a breaker assembly. These surfaces should *never* be touched with a file. The contact faces are plated with hard alloy to withstand the heat of the counterspark. If you file this plating off, the points burn very quickly.

However, you can clean and burnish the faces of the contacts with a sheet of crocus cloth. The grit of crocus cloth is so fine, it only polishes the faces and cleans off residue. Even this, though, is for emergencies, when you cannot install new points. Lay the crocus cloth on any flat surface. Holding the contact face precisely flat on the crocus cloth, rub the contact back and forth lightly. Try to keep the contact face as flat as possible. It does not take much to polish whatever can be polished without damage.

The only trouble you will have installing this type of condenser centers around getting the coil and kill wires into the hole in the contact stem of the condenser. Find a small, round, pointed tool such as a scratch awl or a small nail. Its tip must fit into the hole to hold the compression spring down while you push the wires through the hole, as shown in Fig. 108.

You must slide the holding tool out of the hole as you slide the wires through the hole from the other side. A pair of sharp-nose pliers can help you hold the ends of the wires together while you push them through the hole. The pliers keep the wires from folding up and make the pushing easier.

After you have properly positioned the wires in the condenser stem, clamp the condenser in position. The condenser slides in its mounting so you can set the gap between the contact faces when they are furthest apart.

First turn the crankshaft until the breaker arm contact is pushed as far away from the condenser contact as possible. Then insert the .020 leaf of a clean feeler gauge between the contact faces. Slide the condenser up until both contact faces lightly touch the blade of the feeler gauge. Tighten the clamp screw. Do not let the contacts press very much on the feeler blade or the gap will not end up correct. See Figs. 109 and 110.

The proper gap for the breaker points in small engines is .020 inch. Any error in the gap affects operating efficiency of the engine. Enough error makes the engine hard to start.

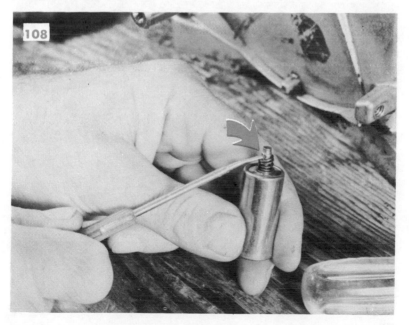

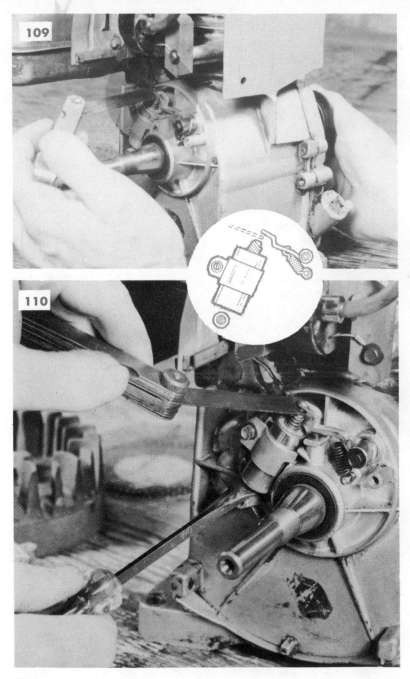

TWO-CYCLE TUNE-UPS

The breaker points assembly in a two-cycle engine does not vary much from the assembly in a four-cycle engine. To replace the points assembly, you must first remove the flywheel. Place a screwdriver or other tool between two fins of the flywheel to hold it as you break the retaining nut loose. Again, remember, be very careful and not break the fins on the flywheel.

This nut also holds the dirt screen and hand-start pulley on the flywheel. To remove the flywheel, use a hammer for tapping and a screwdriver for prying, as previously described. See Figs. 111 and 112. Be sure you replace the nut on the crankshaft and thread it down flush before you hit the end with a hammer, so you do not damage the threads.

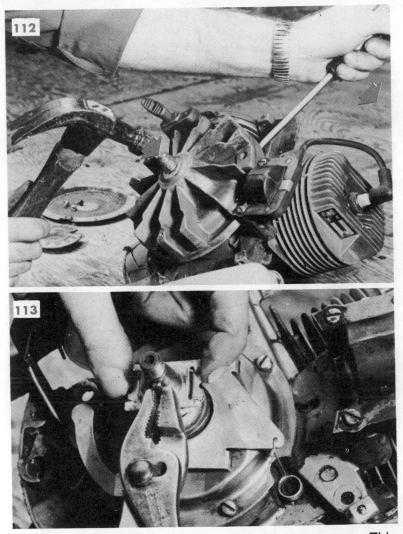

After the flywheel, the governor is next to remove. This particular engine has no standard throttle control like a four-cycle engine. The governor assembly regulates the speed of the engine during operation.

This is a centrifugal governor, with *bob-weights* mounted on the crankshaft under the flywheel. The complex assembly of bob-weights is held in place by the flywheel key in the crankshaft. See Fig. 113.

Remove this key carefully while you hold the bob-weight together with your fingers. If you do not hold the array together, it can come apart in your hands when you slide it from the crankshaft. If the bob-weights fall out of their retainers, carefully reassemble them before you try to replace the assembly on the crankshaft. In fact, it is advisable to reassemble them before you set them aside to proceed with other work; it is too easy to forget how they fit together. See Fig. 114.

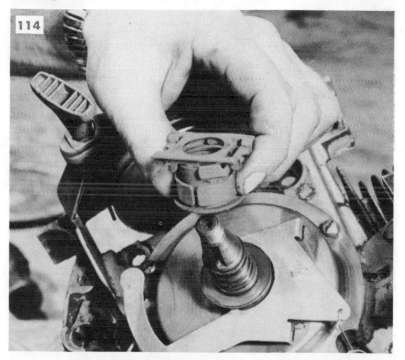

A spring under the bob-weight assembly holds the assembly up against the underside of the flywheel. Remove this spring from the crankshaft. See Fig. 115. Under this spring you will find a large metal control plate. One projection from it sticks out the side of the engine so you can vary the speed of the engine. This control plate pivots on two tabs that fit into slots in the breaker-points cover. A coil spring down the side of the engine supplies tension against which the governor mechanism works. Unhook the spring

with a pair of pliers before you remove the control plate. See Fig. 116. With the spring loose, raise the plate slightly to clear the tabs in the points cover. Slide the plate away from its slide ring that moves up and down on the crankshaft. See Fig. 117.

Three screws secure the points cover over the breaker-points compartment. As you remove these screws, observe how the slots for the governor control plate are oriented. You must put the slots in the same place when you replace

the cover. See Fig. 118. Lifting off the points cover gives easy access to the points assembly, as shown in Fig. 119.

Remove the nut from the terminal where the kill wire and coil wire are connected. Notice that this nut also holds the breaker-arm spring, providing the electrical connection to the movable points contact. Further, you will also notice that the bolt this nut screws onto is the stem of the condenser, in this model engine.

The first wire to come off of the breaker-assembly terminal screw is the kill wire. Push this wire out of your way

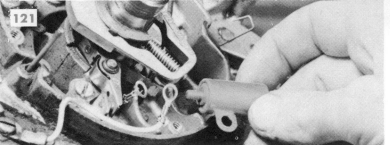

until you replace it on a new breaker assembly. This wire runs from the breaker assembly to a terminal on the engine near the primer button. A tab on the primer shaft touches the wire terminal and shorts out the coil. Then no spark can go to the spark plug, and the engine stops running. See Fig. 120.

The next item you remove is the condenser. It mounts adjacent to the breaker assembly, held in place by one screw. This condenser has a short bolt-thread on the stem. This is the terminal screw for the points assembly. The mounting screw grounds the condenser case to the engine frame. See Fig. 121.

The last segment of the governor control, an automatic advance that alters ignition timing to match the load on the engine, combines with the cam ring on the crankshaft. The cam ring, you recall, pushes the breaker arm to open and close the points.

A pin through the cam ring holds the complete assembly on the crankshaft. A small spring and clip keep the pin in place, while allowing the advance bracket freedom to move. See Figs. 122 and 123.

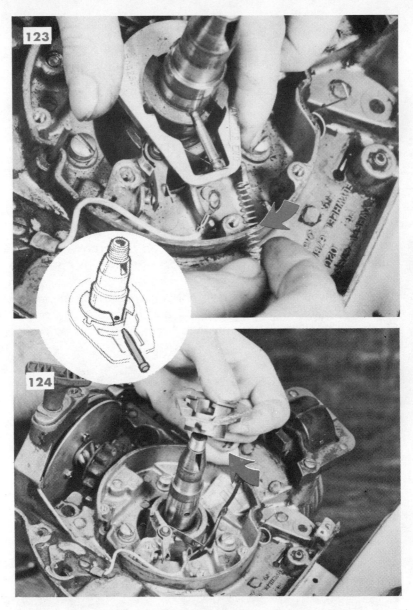

To remove this pin, pull the spring clip off with a pair of sharp-nose pliers. Slide the pin out of the bracket and crankshaft. When the pin is out, the entire advance assembly can be removed from the crankshaft. See Fig. 124.

Removing the time advance mechanism and cam ring clears your path to the breaker-points assembly. Half the breaker assembly mounts rigidly to the base. The other half is the movable arm. Each half has one contact face. The position of the rigid half is adjustable; this is how you set the breaker gap. See Fig. 125.

Installing new points is the reverse of taking out the old ones. Be careful of positioning. Once the new points set is in and tightened down, replace the cam ring and advance mechanism on the crankshaft. Reconnect the coil and kill wires, along with the breaker arm spring, on the condenser

125

terminal. Tighten the nut, and be sure the wire lugs do not touch metal anywhere except the terminal.

Set the breaker gap, as shown in Fig. 126. Turn the crankshaft until the cam opens the points at the widest position. Set the gap to .020 inch with a clean feeler gauge. Maneuver the rigid half of the points set with a screwdriver in the twist-slot. Then tighten the mounting screw. Reassemble the engine: governor plate, flywheel, governor weights, starter, and cover.

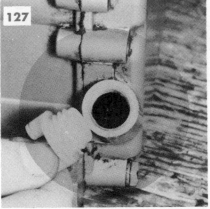

One last item in any tune-up operation: CHECK THE OIL LEVEL in four-cycle engines every time you do any work. The oil level should also be verified before starting, every time the engine is used. The oil should come to just below the plug in the filler hole.

You should change oil every 25 hours of running time. Drain the old oil into a coffee can or soft-drink can. The drain hole is on the side, at the bottom, on a horizontal-shaft engine. See Fig. 127. On a vertical-shaft engine, the drain plug is on the underside of the engine. When you replace the oil, use only a top grade nondetergent oil, SAE 30. Refill the reservoir to just below the filler hole.